GOODS OF THE MIND, LLC

Competitive Mathematics Series

for

Gifted Students in Grades 1 and 2

PRACTICE ARITHMETIC

Cleo Borac, M. Sc.
Silviu Borac, Ph. D.

This edition published in 2013 in the United States of America.

Editing and proofreading: David Borac, B.Mus.
Technical support: Andrei T. Borac, B.A., PBK

Send all inquiries to:

Goods of the Mind, LLC
1138 Grand Teton Dr.
Pacifica
CA, 94044

Competitive Mathematics Series for Gifted Students
Level I (Grades 1 and 2)
Practice Arithmetic

Contents

Contents

FOREWORD

The goal of these booklets is to provide a problem solving training ground starting from the earliest years of a student's mathematical development.

In our experience, we have found that teaching how to solve problems should focus not only on finding correct answers but also on finding better solution strategies. While the correct answer to a problem can typically be obtained in several different ways, not all these ways are equally useful for learning how to solve problems.

The most basic strategy is *brute force*. For example, if a problem asks for the number of ways Lila and Dina can sit on a bench, it is easy to write down all the possibilities: Dina, Lila and Lila, Dina. We arrive at this solution by performing all the possible actions allowed by the problem, leaving nothing to the imagination. For this last reason, this approach is called brute force.

Obviously, if we had to figure out the number of ways 30 people could stand in a line, then brute force would not be as practical, as it would take a prohibitively long time to apply.

Using brute force to obtain the correct answer for a simpler problem is not necessarily a useful learning experience for solving a similar problem that is more complex. Moreover, solving problems in a quantitative manner, assuming that the student can transfer simple strategies to similar but more complex problems, is not an efficient way of learning problem solving.

From this simple example, we see that the goal of *practicing* problem solving is different from the goal of problem solving. While the goal of problem solving is to obtain a correct answer, the goal of practicing problem solving is to acquire the ability to develop strategies, generate ideas, and combine approaches that are powerful enough to solve the problem at hand as well as future similar problems.

While brute force is not a useless strategy, it is not a key that opens every door. Nevertheless, there are problems where brute force can be a useful tool. For instance, brute force can be used as a first step in solving a complex problem: a smaller scale example can be approached using brute force to help the problem solver understand the mechanics of the problem and generate ideas for solving the larger case.

All too often, we encounter students who can quickly solve simple problems by applying brute force and who become frustrated when the solving methods they have been employing successfully for years become inefficient once problems increase in complexity. Often, neither the student nor the parent has a clear understanding of why the student has stagnated at a certain level. When the only arrows in the quiver are guess-and-check and brute force, the ability to take down larger game is limited.

Our series of books aims to address this tendency to continue on the beaten path - which usually generates so much praise for the gifted student in the early years of schooling - by offering a challenging set of questions meant to build up an understanding of the problem solving process. Solving problems should never be easy! To be useful, to represent actual training, problem solving should be challenging. There should always be a sense of difficulty, otherwise there is no elation upon finding the solution.

Indeed, practicing problem solving is important and useful only as a means of learning how to develop better strategies. We must constantly learn and invent new strategies while questioning the limitations of the strategies we are using. Obtaining the correct answer is only the natural outcome of having applied a strategy that worked for a particular problem in the time available to solve it. Obtaining the wrong answer is not necessarily a bad outcome; it provides insight into the fallacies of the method used or into the errors of execution that may have occured. As long as students manifest an interest in figuring out strategies, the process of problem solving should be rewarding in itself.

Sitting and thinking in a focused manner is difficult to train, particularly since the modern lifestyle is not conducive to adopting open-ended activities. This is why we would like to encourage parents to pull back from a quantitative approach to mathematical education based on repetition, number of completed pages, and the number of correct answers. Instead, open up the time boundaries that are dedicated to math, adopt math as a game played in the family, initiate a math dialogue, and let the student take his or her time to think up clever solutions.

Figuring out strategies is much more of a game than the mechanical repetition of stepwise problem solving recipes that textbooks so profusely provide, in order to "make math easy." Mathematics is not meant to be easy; it is meant to be interesting.

Solving a problem in different ways is a good way of comparing the merits of each method - another reason for not making the correct answer the primary goal of the activity. Which method is more labor intensive, takes more time or is more prone to execution errors? These are questions that must be part of the problem solving process.

In the end, it is not the quantity of problems solved, the level of theory absorbed, or the number of solutions offered in ready-made form by so many courses and camps, but the willingness to ask questions, understand and explore limitations, and derive new information from scratch, that are the cornerstones of a sound training for problem solvers.

These booklets are not a complete guide to the problem solving universe, but they are meant to help parents and educators work in the direction that, aside from being the most efficient, is the more interesting and rewarding one.

The series is designed for mathematically gifted students. Each book addresses an age range as some students will be ready for this content earlier, others later. If a topic seems too difficult, simply try it again in a couple of months.

DIGITS AND NUMBERS

Digits are symbols we use to write out numbers. In the decimal system there are 10 digits available: 0, 1, 2, 3, 4, 5, 6, 7, 8, and 9.

Numbers are written using digits. For example, 23 is a 2-digit number, 6 is a one-digit number, and 50031 is a 5-digit number.

0, 2, 4, 6, and 8 are *even* digits.

1, 3, 5, 7, and 9 are *odd* digits.

The *digit sum* of a number is simply what we get if we add all the digits.

Because digits range from 0 to 9, interesting observations can be made about the size of the carryover when we add or multiply numbers.

For example, the *only* carryover possible when adding two digits is 1, because two digits add up to at most 18. Even if two 9s are added and there is a carryover of 1 from the previous place value, the carryover will still not exceed 1.

One can find the largest possible carryover when adding three or more digits using a similar reasoning.

Experiment

1. Write the smallest odd numbers with 4, 5, and 6 digits.

2. Write the smallest 3-digit number with odd digits that are all different.

3. Write the largest 4-digit number with a digit sum equal to 3.

4. Write a 6 digit number so that the product of its digits is equal to 1.

5. Write the smallest 4-digit number that can be written using the digits 0, 3, and 7 at least one time each.

6. Write the largest number that contains each odd digit once.

7. Write the smallest number that contains all the even digits.

Answers:

1. 1001, 10001, and 100001

2. 135

3. 3000

4. 111111

5. 3007

6. 97531

7. 20468

Practice One

Exercise 1

Make a list of all the numbers from 1 to 30 that have only even digits.

Exercise 2

What is the largest even 3-digit number?

Exercise 3

What is the largest 3-digit number with all digits even?

Exercise 4

The figure below is painted on the side of a truck. If the truck has stopped beside a lake, how many digits can be seen in its reflection in the water?

0123456789

Exercise 5

We have the following 2 cards:

86   **69**

How many different numbers can we make using them? The cards are square and may be rotated, but they are opaque (not see-through.)

Exercise 6

If we add two digits, what is the largest carryover we can get?

(**A**) 1

(**B**) 2

(**C**) 9

(**D**) any number

Exercise 7

If we add three digits, what is the largest carryover we can get?

Exercise 8

Lila adds two digits and gets a carryover of 1. One of the digits is smaller than 5. The other digit is:

(**A**) larger than or equal to 5

(**B**) larger than 5

(**C**) smaller than 5

(**D**) any digit

Exercise 9

Dina tries to write the number 121212121212 as the sum of two numbers that only have digits of 1 and 0. Can you help her?

Exercise 10

Which is the larger sum?

111111	123456
222222	234561
333333	345612
444444	456123
555555	561234
666666	612345
———— +	———— +

Exercise 11

If you count backwards from 999 to 1, at which count will the digit in the middle change for the first time?

Exercise 12

How many three digit numbers can be formed using only the digits 7 and 0?

Exercise 13

Make a list of the digit sums for all the numbers between 11 and 20.

Exercise 14

Lila made a list of all the numbers smaller than 100 that have a digit sum of 16. How many different digits did she use while writing the list?

Exercise 15

How many 5 digit numbers have a digit sum of 2?

Exercise 16

Dina has written down a 6 digit number. Lila reverses the digits of this number. Then, the two girls compare their numbers and find that they are identical. What is the largest number of different digits Dina's number could have?

Exercise 17

Lila asks Dina: "For how many even 3-digit numbers does the digit sum equal 26?" Dina finds a number. Can you help her find more?

Exercise 18

The number 333 is written using only the digit 3. In how many different ways can we write it as a sum of two positive numbers that are written using only the digits 3 and zero?

Exercise 19

Dina made a list of all the numbers written by repeating the same digit that are greater than 405678 and smaller than 999110. Lila counted the numbers on Dina's list. How many numbers did Lila count?

Exercise 20

Dina thinks of a number and Lila thinks of a number. When they add their numbers, the sum is even. If they subtract the smaller number from the larger number, is the difference even or odd?

Exercise 21

If Lila gave Dina 5 beads, they would both have the same number of beads. What is the difference between the number of beads Lila has and the number of beads Dina has?

Exercise 22

If Lila gave Dina 5 beads, they would both have the same number of beads. Which of the following could not be the total number of beads they have?

(A) 21 beads

(B) 22 beads

(C) 24 beads

(D) 104 beads

Exercise 23

What is the largest possible digit sum for a 3-digit number in which all the digits are different?

Exercise 24

A computer begins counting by ones starting from 13579. What is the next number with odd digits that are all different?

Exercise 25

Arbax, the Dalmatian, has 16 bones hidden in 5 different caches. Arbax thinks there is an odd number of bones in each cache. Is Arbax right?

CRYPTARITHMS

A *cryptarithm* is a mathematical riddle. It consists of a simple operation, such as an addition or a multiplication, in which some or all the digits are replaced by symbols (*encrypted*).

The rules of the cryptarithm are often part of the problem statement and are generally, but not always, as follows:

- different symbols represent different digits,
- the same symbol represents the same digit.

Cryptarithms can be solved using a variety of techniques. Among these techniques, the following are the most important:

- no number ever starts with a 0,
- digits range from 0 to 9 only,
- the carryover is limited to certain values, based on the operation and the number of digits,
- digits already discovered can be ruled out for the remaining symbols,
- the parity of the digits (even or odd) is sometimes predictable,
- if it is possible to bracket the values of a symbol within narrow limits (for example, we conclude that some symbol can only be 1 or 3) it is usually a good idea to try out each value in turn.

Experiment

Try this very simple cryptarithm:

$$\heartsuit\spadesuit + \spadesuit = \clubsuit\diamondsuit\diamondsuit$$

Because the result is a three digit number, its leftmost digit can only be a 1 that has been carried over from the previous place value.

This carryover can only be obtained if the digit marked as a $\heartsuit$ is a 9 and there is a carryover of 1 from the previous place value.

The $\clubsuit\diamondsuit$, therefore, represents the number 10.

Adding the two $\spadesuit$ symbols must produce a carryover and a last digit of zero. Therefore, the $\spadesuit$ must be a 5, and the addition is:

$$95 + 5 = 100$$

When solving cryptarithms, it is important to work logically and not by trial-and-error. The student should work from clue to clue until the problem is completely decrypted.

Practice Two

Exercise 1

In the following figure, different shapes represent different digits, and the same shape represents the same digit. Which digit does the square represent?

Exercise 2

In the following sum, different letters represent different digits:

$$A + A + A + A + A = B$$

Find the digits!

Exercise 3

If A and B are digits and $A + A = B$, how many different values can A have?

Exercise 4

Two different digits cannot have a difference of:

(A) 0

(B) 3

(C) 5

(D) 9

Exercise 5

In the following figure, different shapes represent different digits and the same shape represents the same digit. Which digit does the square represent?

Exercise 6

Two numbers are encoded as AC and BC, where different letters represent different digits and the same letter always represents the same digit. Their sum could be any of the following, except:

(A) 100

(B) 102

(C) 103

(D) 104

Exercise 7

Each symbol represents a digit from 0 to 9. Different symbols represent different digits. Find out what the digits are. How many such additions are there?

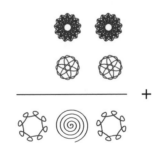

Exercise 8 Lila has found a solution to the following cryptarithm. Arbax, the Dalmatian, is also working hard to find a solution. Can he find a solution different from Lila's?

$$
\begin{array}{r}
\textbf{XOX} \\
\textbf{XOX} \\
\hline
\textbf{OHH}
\end{array} \; +
$$

Exercise 9 Which digits do the symbols represent?

$$
\begin{array}{ccc}
\square & \square & \square \\
\bigcirc & \bigcirc & \bigcirc \\
\hline
\end{array}
$$

$$
\begin{array}{cccc}
\bigstar & \bigcirc & \bigcirc & \bigstar
\end{array} \; +
$$

Exercise 10 Each letter represents a different digit. The same letter is always the same digit. Find the digit that corresponds to each letter:

$$
\begin{array}{cccc}
 & & & D \\
 & & D & D \\
 & D & D & D \\
\hline
A & A & B & C
\end{array} \; +
$$

Exercise 11 In the following examples, O represents an odd digit and E

18

represents an even digit. Which example is impossible?

$$\frac{\begin{array}{cc} O & E \\ E & O \end{array}}{\begin{array}{cc} E & O \end{array}} + \text{(A)} \qquad \frac{\begin{array}{cc} E & E \\ O & O \end{array}}{\begin{array}{cc} O & O \end{array}} + \text{(C)}$$

$$\frac{\begin{array}{cc} O & E \\ E & E \end{array}}{\begin{array}{cc} O & O \end{array}} + \text{(B)} \qquad \frac{\begin{array}{cc} O & E \\ O & O \end{array}}{\begin{array}{cc} E & O \end{array}} + \text{(D)}$$

EQUATIONS

Solving problems using *equations* can be introduced at this age level by using concrete representations for the quantities involved. To solve problems using a concrete technique, use the following steps:

- read the problem carefully,
- identify the quantity that is the most basic and model it using a rectangle,
- use the same rectangle for the same amount,
- observe how the rectangles must be combined,
- figure out which arithmetic operation models the required combination.

Example

Dina had 4 more dinosaurs than Lila. Lila received 16 more dinosaurs for her birthday and now she has twice as many dinosaurs as Dina. How many dinosaurs did Lila have before her birthday?

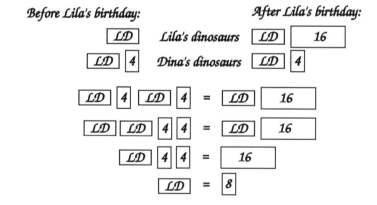

Example

A baker has made twice as many baguettes as loaves and half as many rolls as loaves. If he now has 140 products up for sale, how many baguettes has he made?

Find the product with the smallest amount. It's the rolls, right? Make a rectangle that represents the number of rolls:

Rolls

Read the problem again. There are twice as many loaves as rolls.

Rolls	*Rolls*	=	*Loaves*

Read the problem again. There are twice as many baguettes as loaves.

Loaves	*Loaves*	=	*Baguettes*

Now make a diagram with all the products using the number of rolls as a unit:

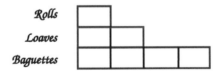

There are 7 rectangles. If there are 140 products in total, each rectangle represents 20 products. Therefore, he baked 20 rolls, 40 loaves, and 80 baguettes.

PRACTICE THREE

Exercise 1

Dina walked 6 steps up from the bottom of a staircase while Lila walked 5 steps down from the top. They met on the same step. How many steps does the staircase have?

Exercise 2

When Arbax, the Dalmatian, celebrates his 7th birthday, Lila will be 16. Today, it is Arbax's birthday and Lila is 10. How old is Arbax?

Exercise 3

Dina's mother is cooking stuffed mushrooms. For each mushroom, she chops one tomato and two scallions. She fills a tray with 6 stuffed mushrooms. How many vegetables has she used in total?

Exercise 4

There are four chairs at the breakfast counter in Lila's kitchen. At breakfast, Dina, Lila, and their parents sit down for their meal. How many legs can one count?

Exercise 5

On a good spring day, one can see deer and hares from Dina's porch. Today, Dina saw as many hare's ears as deer's tails. If Dina saw 7 hares, how many deer did she see?

Exercise 6

Lila gave Amira 12 magic wands. Amira gave Dina 8 magnets. Dina gave Lila 14 crystals. Which girl experienced the largest difference in her total number of toys? Was it an increase or a decrease?

Exercise 7

Arbax, the Dalmatian, had stashed away 25 bones. Some cats discovered his cache and managed to run away with 2 bones each as Arbax chased them away. Arbax is a smart dog who can count. He counted the bones and found that there were 17 left. How many cats were there?

Exercise 8

Lila has 18 new blue pens. Amira proposes an exchange: she will give Lila one fluorescent pen for every 7 blue pens. Lila wants 4 fluorescent pens. Amira will accept a scented eraser instead of 2 blue pens to make up for the difference. How many erasers does Lila have to give Amira?

Exercise 9

Which two consecutive months have 62 days in total?

Exercise 10

Five rats, two squirrels, and one mongoose decide to start a small business. In one day, each rat can bring in 10 clients and each squirrel can bring in 6 clients. The mongoose is in charge of the supply chain and has to provide 4 dollars worth of merchandise per client. How many days will it take the mongoose to spend 128 dollars?

Exercise 11

Each day, Penelope weaves 6 feet of fabric. Each night, she unravels 5 feet of fabric. How many days does it take her to weave a 12 ft long coverlet?

Exercise 12

An unusual weather pattern started on a Monday. Each day it rained for three hours as follows: the first day it rained from midnight to 3 am, the next day it rained from 3 am to 6 am, and so on. What day of the week was it when the rain started at midnight for the second time?

Exercise 13

Five bears foraged together in a forest. They decided to take turns to rest as follows: one of them would rest while the other four foraged. If each bear rested for 6 hours, how many hours did each bear spend foraging?

Exercise 14

It's dance recital day! Lila had her performance at 3 pm and arrived 2 hours ahead of time for rehearsal. Dina had her performance at 2 pm and arrived 3 hours ahead of time, when the hall opened, for rehearsal. Each performance was one hour long and the hall closed after Lila's performance. For how many hours was the hall open?

Exercise 15

Lila's train, which was supposed to arrive at 3:45 pm, arrived 20 minutes late. As a result, Lila missed the 3:50 bus and had to wait 25 minutes for the next bus. If the bus ride home takes 15 minutes, how late was Lila compared to her usual arrival time?

Exercise 16

Dina has a Book of Magic that is numbered backwards. If she is now on page 201 and the book has 453 pages, how many pages has she actually read?

Exercise 17

Cornelia, the shepherd, has 70 animals: sheep, hens, and cows. There are as many sheep's legs as hen's legs. There are as many cow's tails as hen's legs. How many cows are there?

Exercise 18

How many pounds of cabbage at 2 dollars a pound weigh the same as 5 pounds of onions at 1 dollar a pound?

Exercise 19

Lila had to help her mother mail some reports. She used a staple for every 3 sheets of paper. As she stapled, 20 staples broke and needed replacement. How many staples did she need if she had to staple 480 sheets in total?

Exercise 20

Ali and Baba are exploring the cave of the forty thieves. The entrance is locked, but Baba has a magic key that unlocks any lock. From the entrance, 5 corridors branch out, each of them locked. Each corridor has 3 rooms on one side and 3 rooms on the other side, all locked. How many of the rooms will Ali and Baba be able to explore if the magic key loses its power after 20 uses?

MISCELLANEOUS PRACTICE

Exercise 1

Jerome had twelve properly outfitted boats ready to be rented out at his beach sports shop. Each boat is required to have two paddles and four lifesavers. A storm swept away five paddles and eight lifesavers from the shop. How many operational boats did he have remaining?

Exercise 2

Lila is learning how to cook. She has a recipe for cookies that calls for 2 cups of flour, 3 sticks of butter, one cup of chocolate chips, and three eggs. She uses this recipe and makes 26 cookies. How many cups of flour should she use next time if the wants to bake 39 cookies?

Exercise 3

Dina, Lila, and Amira have 5 plush toys. Can you help them share the toys so that each of them has a different number of toys? (Each girl must have at least one toy.)

Exercise 4

Ali discovered a box filled with gold coins in the thieves' lair. Ali managed to stuff half of the coins in his pockets and Baba managed to stuff half of the remaining coins in his pockets before they heard noises from outside and slipped out of the cave. The returning thieves found 30 coins in the box. How many coins did Ali and Baba steal?

Exercise 5

If the letter **O** represents an odd digit, and the letter **E** represents an even digit, which of the following equalities is impossible?

(A) $O + O = E$

(B) $OO - OO = EE$

(C) $OO - O = EE$

(D) $EE + O = EO$

(E) $E + O = EO$

Exercise 6

Dina's grandmother has 5 blue cups, 2 red cups, 3 red saucers, 2 blue saucers, and 2 green saucers. How many cup and saucer pairs can she make for which the color of the saucer does not match the color of the cup?

Exercise 7

Tony, the car mechanic, wants to have equal amounts of coolant fluid in two containers. He has 4 liters of coolant in one container and 6 liters in the other. How many liters of coolant should he transfer from the first container into the second one?

Exercise 8

Dina and Amira are standing in line at the baker's shop. There are four people in front of Dina and four people between Dina and Amira. By the time Dina places her order, Amira will be:

(A) third in line

(B) fourth in line

(C) fifth in line

(D) sixth in line

(E) eighth in line

Exercise 9

Lila and Dina are playing "Little Romans." They have to dress in togas, speak Latin, and solve the following:

$$MCM - MC =$$

Exercise 10

Help Dina and Lila solve their Roman homework by providing results expressed as Roman numerals:

(a) $CM + MC =$

(b) $MCCC + DCCC =$

(c) $XI + IX =$

(d) $MC - CM =$

(e) $XI - IX =$

(f) $XXIV + XXVI =$

(g) $XXVI - XXIV =$

(h) $LX - XL =$

(i) $MMM - CCC =$

(j) $MCL - CML =$

(k) $CXX - LXX =$

(l) $CCC - XXX =$

Exercise 11

1. How many two digit numbers have identical digits?

2. How many three digit numbers have identical digits?

3. How many four digit numbers have identical digits?

4. How many one hundred digit numbers have identical digits?

Exercise 12

Lila has some cards with the digit 6 on them. By turning a card upside down, she can get a card with the digit 9 on it. Lila forms two digit numbers with these cards. What is the sum of all the different numbers she can make?

Exercise 13

In the figure, the square and the circle represent different digits. What is the difference between the largest and the smallest digit values the triangle can represent?

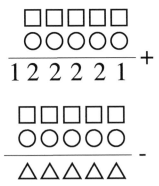

Exercise 14

The figure represents a correct addition:

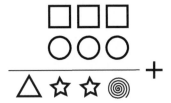

Different shapes represent different digits and the same shape always represents the same digit. Which of the following values can the spiral not have? Check all that apply.

(A) 0

(B) 1

(C) 7

(D) 8

(E) 9

Exercise 15

In the following list, how many numbers are odd?

$$4, \ 10, \ 151, \ 200, \ 0, \ 45, \ 736$$

Exercise 16

Find the following numbers:

1. the largest three digit number with all different odd digits
2. the smallest three digit number with all different odd digits
3. the largest three digit number with all different even digits
4. the smallest three digit number with all different even digits

Exercise 17

The following sequence is built according to a pattern. Help Dina add three more terms to it!

$$79, \ 69, \ 60, \ 52, \cdots$$

Exercise 18

The sum $\Diamond + \Diamond$, where $\Diamond$ represents a digit, is:

(A) odd

(B) even

(C) impossible to determine

(D) neither even nor odd

Exercise 19

Amira has to make 3 different prizes for the swim meet. She has 5 toys to use as prizes, but each prize has to consist of a different number of toys. Can you explain why she is a bit puzzled?

Exercise 20

The largest of five consecutive odd numbers is 19. How many of these numbers are larger than 14?

Exercise 21

Which number does the star represent?

$$\bigcirc + \bigcirc + \bigcirc = 60$$

$$\circledcirc + \circledcirc = \bigcirc$$

$$\bigcirc + \circledcirc = \bigstar$$

Exercise 22

Dina wrote four consecutive odd numbers on a piece of paper. Arbax played with it and pierced it with his fangs. Now some of the digits are no longer legible. Can you figure out the missing digits?

$$\_\,3\,1\,\_$$

$$\_\,3\,\_\,9$$

$$5\,\_\,\_\,\_$$

$$\_\,\_\,\_\,\_$$

Exercise 23

Lila had 2 blue cubes worth 5 points each, 2 green cubes worth 3 points each, and 2 yellow cubes worth 6 points each. She placed them in a bag and asked Dina to pick 3 cubes without looking.

1. What is the largest odd total score Dina can obtain?
2. What is the lowest even total score Dina can obtain?

Exercise 24

Dina's mother organized a party for the children of her coworkers. She placed some boxes of candy on a table. Each box contained 20 candies. After the party, Lila noticed a funny thing had happened: one box was left untouched, from another only one candy had been removed, from another two candies were missing, from another three candies, and so on. The final box was completely empty. How many boxes were there in total?

Exercise 25

Lila and Amira are selling 40 of their old toys at a yard sale. Instead of selling each toy for 90 cents, they decided the sale would go quicker if they offered a "buy 3 toys and get one toy free" deal. To make the same amount of money in total, how much do they have to sell each toy for? (Assume they sell all the toys in either case.)

Exercise 26

Lila and Dina are trading toys. They agree that one doll should be worth 3 magnets and that one magnet should be worth two UV-beads. Lila has lots of UV-beads. If Lila wants to trade UV-beads for a doll and a magnet from Dina, how many UV-beads should she give Dina?

Exercise 27

Amira has stored 759 songs in her collection of favorites. How many more songs does she have to store for the number of songs to be the next number with three different odd digits?

Exercise 28

Dina and Lila are allowed to remove 3 digits from the number 13904521 but are not allowed to change the places of the remaining digits. Dina wants to obtain the largest possible number formed with the remaining digits, while Lila wants to obtain the smallest. What is the difference between Dina's number and Lila's number?

Exercise 29

Max, the baker, sells baguettes for 1.20 dollars each and focaccias for 3.40 dollars each. What is the smallest number of baguettes and foccacias he must sell in order to get an amount of money that can be written as whole dollars, no cents.

Exercise 30

For each puffin that flies off from a cliff, three puffins land on the cliff. There were 1000 puffins on the cliff and 20 of them flew off. How many puffins are there now on the cliff?

SOLUTIONS TO PRACTICE ONE

Exercise 1

Make a list of all the numbers from 1 to 30 that have only even digits.

Solution 1

2, 4, 6, 8, 20, 22, 24, 26, and 28

Exercise 2

What is the largest even 3-digit number?

Solution 2

998

Exercise 3

What is the largest 3-digit number with all digits even?

Solution 3

888

Exercise 4

A figure is painted on the side of a truck. If the truck has stopped beside a lake, how many digits can be seen in its reflection in the water?

Solution 4

Four of the ten digits: 0, 1, 3, and 8.

0123456789

0123456789

Exercise 5

How many different numbers can we make using the cards?

Solution 5

The card 86 turns into the card 98 when rotated half a circle.
The card 69, on the other hand, remains the same when rotated half a circle.

We can make 7 numbers:

Solution 6

If we add two digits, the largest carryover we can get is 1 because the largest digit is 9.

Solution 7

If we add three digits, the largest carryover is 2, since $9 + 9 + 9 = 27$.

Solution 8

Lila adds two digits and gets a carryover of 1. If one of the digits is smaller than 5, the other digit is necessarily larger than 5. Two digits that are both smaller than 5 do not produce a carryover when added.

Solution 9

Dina tries to write the number 121212121212 as the sum of two numbers that only have digits of 1 and 0. Can you help her?

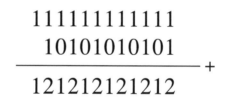

Solution 10

Both sums have the same value. Take, for example, the digit 1. It appears one time in the hundred thousands place, one time in the ten thousands, one time in the thousands, one time in the hundreds, one time in the tens, and one time in the units in *both* sums. The same is true about each of the other digits. Notice that we did not need to compute the actual values of the sums.

Exercise 11

If you count backwards from 999 to 1, at which count will the digit in the middle change for the first time?

Solution 11

At the 11$^\text{th}$ count:

1-st count	999
2-nd count	998
3-rd count	997
4-th count	996
5-th count	995
6-th count	994
7-th count	993
8-th count	992
9-th count	991
10-th count	990
11-th count	989

Exercise 12

How many three digit numbers can be formed using only the digits 7 and 0?

Solution 12

Strategic Approach: No number can start with 0. Therefore, we can only have numbers that start with 7. There is only one choice for the first digit. There are 2 choices for the second digit and 2 choices for the third digit. The total number of possibilities is $2 \times 2 = 4$.

Brute Force Approach: Make a list of the numbers:

$$\{700, 707, 770, 777\}$$

Exercise 13

Make a list of the digit sums for all the numbers between 11 and 20.

Solution 13

Strategic Approach: 11 has a digit sum of 2. As we count up from 11 to 19 only the last digit changes, each time by 1. So far, we will get all the sums between 2 and 10. When we reach 20 the digit sum will

37

once again be 2. Therefore, the sums are all the consecutive integers from 2 to 10.

Brute Force Approach: Make a list of the digit sums:

11	2
12	3
13	4
14	5
15	6
16	7
17	8
18	9
19	10
20	2

Exercise 14

Lila made a list of all the numbers smaller than 100 that have a digit sum of 16. How many different digits did she use while writing the list?

Solution 14 16 is a pretty large sum for a two digit number. Only a few numbers can have such a large digit sum:

$$\{79, \ 88, \ 97\}$$

Lila used 3 different digits to write the numbers in her list.

Exercise 15

How many 5 digit numbers have a digit sum of 2?

Solution 15

There are 4 such numbers. Since no number can start with 0, there is only one number with a digit of 2 and 4 digits of zero: 20000. Then, there are numbers that have two digits of one and three digits of zero. These must all start with the digit 1; there are four other positions remaining for the other digit of 1:

38

$$\{10001, \ 10010, \ 10100, \ 11000\}$$

Exercise 16

Dina has written down a 6 digit number. Lila reverses the digits of this number. Then, the two girls compare their numbers and find that they are identical. What is the largest number of different digits Dina's number could have?

Solution 16

For the two numbers to be equal, Dina must have written down a *palindrome*. A 6-digit palindrome has at most three different digits. Example:

$$123321$$

Of course, it is possible to satisfy the condition with even fewer distinct digits, such as:

$$121121 \text{ or } 111111$$

Exercise 17

Lila asks Dina: "For how many even 3-digit numbers does the digit sum equal 26?" Dina finds a number. Can you help her find more?

Solution 17

No. Dina has found the only number that meets Lila's requirements: 998.

Exercise 18

The number 333 is written using only the digit 3. In how many ways can we write it as a sum of two positive numbers that are written using only the digits 3 and zero?

Solution 18

In three different ways:

$$333 = 300 + 33$$
$$333 = 303 + 30$$
$$333 = 330 + 3$$

Because addition is *commutative*, ways in which the same terms are added in a different order are not distinct from these.

Exercise 19

Dina made a list of all the numbers written by repeating the same digit that are greater than 405678 and smaller than 999110. Lila counted the numbers on Dina's list. How many numbers did Lila count?

Solution 19

Dina's list looks like this:

$$\{444444, 555555, 666666, 777777, 888888\}$$

Lila found that the list contains 5 numbers.

Exercise 20

Dina thinks of a number and Lila thinks of a number. When they add their numbers, the sum is even. If they subtract the smaller number from the larger number, is the difference even or odd?

Solution 20

If two numbers have an even sum, they also have an even difference.

Exercise 21

Lila could give Dina 5 beads in order for them to have the same number of beads. What is the difference between the number of beads Lila has and the number of beads Dina has?

Solution 21

Dina has 10 beads more than Lila.

Solution 22

If the total number of beads can be divided into two equal parts, then it must be even. The impossible choice is 21 beads.

Exercise 23

What is the largest possible digit sum for a 3-digit number in which all the digits are different?

Solution 23

Choose the three largest digits: 9, 8, and 7. The largest digit sum is $7 + 8 + 9 = 24$.

Exercise 24

A computer begins counting by ones starting from 13579. What is the next number with odd digits that are all different?

Solution 24

13597

Exercise 25

Arbax, the Dalmatian, has 16 bones hidden in 5 different caches. Arbax thinks there is an odd number of bones in each cache. Is Arbax right?

Solution 25

Arbax cannot be right. If there is an odd number of bones in each cache and there is an odd number of caches, then the total number of bones must be odd. 16, however, is even.

$$O + O + O + O + O = E + E + O = E + O = O$$

Note that we do not need to assign specific numbers of bones to the caches.

SOLUTIONS TO PRACTICE TWO

Exercise 1

In the following figure, different shapes represent different digits, and the same shape represents the same digit. Which digit does the square represent?

Solution 1

The first number has two identical digits and is smaller than 42. Also, since we add only a single digit to obtain 42, the number must be equal to 33. The square represents the digit 9 ($42 - 33 = 9$).

Exercise 2

In the following sum, different letters represent different digits:

$$A + A + A + A + A = B$$

Find the digits!

Solution 2

Only $A = 1$ and $B = 5$ are possible. If $A = 0$ then $B = 0$ and this does not fulfill the requirement that they be different. Other values for A produce a carryover and result in a sum that is no longer a one-digit number.

Exercise 3

If A and B are digits and $A + A = B$, how many different values can A have?

Solution 3

Since the result of the addition is a single digit, there is no carryover. A cannot be larger than 4. Also, since **B** is a digit different from **A**, **A** cannot be zero. Therefore, **A** can be either 1, 2, 3, or 4. There are 4 possible values.

Exercise 4

Two different digits cannot have a difference of:

(A) 0

(B) 3

(C) 5

(D) 9

Solution 4

Two different digits cannot have a difference of 0. The other differences listed are possible.

Exercise 5

In the following figure, different shapes represent different digits and the same shape represents the same digit. Which digit does the square represent?

Solution 5

The square represents a single digit and cannot be greater than 9.

The number formed of two circles, therefore, cannot be greater than 77. Since the circles represent the same digit, the only solution that satisfies the condition is:

$$77 - 9 = 68$$

The square represents the digit 9.

Exercise 6

Two numbers are encoded as AC and BC, where different letters represent different digits and the same letter always represents the same digit. Their sum could be any of the following, except:

(A) 100

(B) 102

(C) 103

(D) 104

Solution 6

Since the last digit of both numbers is the same, the sum must end with an even digit. The only answer choice that does not end with an even digit is (C).

Exercise 7

Each symbol represents a digit from 0 to 9. Different symbols represent different digits. Find out what the digits are. How many such additions are there?

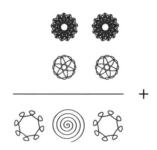

Solution 7

The two tens digits cannot add up to more than 18. Even if there is a carryover from the previous place value, the sum in the tens place cannot exceed 19. Therefore, the hundreds digit of the result must be 1. The units digit of the result has the same symbol, therefore it is also 1. Digits that add up to a number with the last digit 1 and that produce a carryover are: $5+6$, $4+7$, $3+8$, and $2+9$. Therefore, there are four possible solutions:

$$55 + 66 = 121$$
$$44 + 77 = 121$$
$$33 + 88 = 121$$
$$22 + 99 = 121$$

Exercise 8 Lila has found a solution to the following cryptarithm. Arbax, the Dalmatian, is also working hard to find a solution. Can he find a solution different from Lila's?

$$
\begin{array}{r}
\mathbf{XOX} \\
\mathbf{XOX} \\
\hline
\mathbf{OHH}
\end{array} +
$$

Solution 8

When we add two identical digits the sum must be *even*. Therefore, **H** must represent an even digit. Since we have $\mathbf{X} + \mathbf{X} = \mathbf{H}$ in the units place and $\mathbf{X} + \mathbf{X} = \mathbf{O}$ in the hundreds place, one of these two sums must include a carryover. There can be no carryover in the units place, therefore we must have: $\mathbf{X} + \mathbf{X} + 1 = \mathbf{O}$ in the hundreds place. Hence, **O** must be odd.

Since the result ends in two identical digits (**HH**), they must come from adding the two different pairs of digits that produce the same last even digit.

- $0 + 0 = 0$ (no carryover) or $5 + 5 = 10$ (carryover 1)
- $1 + 1 = 2$ (no carryover) or $6 + 6 = 12$ (carryover 1)
- $2 + 2 = 4$ (no carryover) or $7 + 7 = 14$ (carryover 1)
- $3 + 3 = 6$ (no carryover) or $8 + 8 = 16$ (carryover 1)
- $4 + 4 = 8$ (no carryover) or $9 + 9 = 18$ (carryover 1)

Since **O** is odd and **O** + **O** produces a carryover of 1, **O** can be either 5, 7, or 9. 5 is not possible because, in that case, **X** = 0 and 0 cannot be the first digit of a number. There remain two possible choices for **O**:

$$
\begin{array}{r}
272 \\
272 \\
\hline
544
\end{array} +
\qquad
\begin{array}{r}
494 \\
494 \\
\hline
988
\end{array} +
$$

Of these, only the one on the right satisfies the cryptarithm. Arbax cannot find a solution different from Lila's!

Exercise 9

Which digits do the symbols represent?

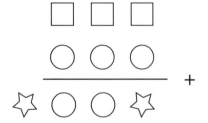

Solution 9

Since the sum of the square and the circle produce either a circle or a star, we conclude that the two digits produce a carryover when added. Since the carryover from adding two digits cannot exceed 1, the star must represent the digit 1 and the circle must be larger than the star by 1. Since the star is 1, the circle must be 2. 2 does not produce a carryover when added to any digit except 9 or 8. In this case, the square represents the digit 9. This is the solution:

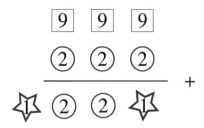

Exercise 10

Find the digit that corresponds to each letter:

$$
\begin{array}{cccc}
 & & & D \\
 & & D & D \\
 & D & D & D \\
\hline
A & A & B & C
\end{array} +
$$

Solution 10

A can only represent 1 because there cannot be a larger carryover from the hundreds place. **D** must be a digit that, when added to some carryover (which can be 1 or 2) produces a last digit of 1. Only 9 satisfies. Therefore, we must have:

$$
\begin{array}{ccccc}
 & & & & 9 \\
 & & & 9 & 9 \\
 & & 9 & 9 & 9 \\
\hline
 & 1 & 1 & 0 & 7
\end{array} +
$$

Exercise 11

In the following examples, O represents an odd digit and E represents an even digit. Which example is impossible?

$$
\begin{array}{cc}
O & E \\
E & O \\
\hline
E & O
\end{array} + \text{(A)} \qquad
\begin{array}{cc}
E & E \\
O & O \\
\hline
O & O
\end{array} + \text{(C)}
$$

$$
\begin{array}{cc}
O & E \\
E & E \\
\hline
O & O
\end{array} + \text{(B)} \qquad
\begin{array}{cc}
O & E \\
O & O \\
\hline
E & O
\end{array} + \text{(D)}
$$

Solution 11

The impossible example is (B), because it shows two even units digits that produce an odd sum. Since there can be no carryover from a previous place value, this is not possible. In the other examples, a possible carryover of 1 from the units place to the tens place can change the parity of the digit accordingly. Here are some possible examples for each type of operation:

$$\frac{\begin{array}{cc} 3 & 8 \\ 2 & 5 \end{array}}{\begin{array}{cc} 6 & 3 \end{array}} + \text{(A)}$$

$$\frac{\begin{array}{cc} 4 & 4 \\ 3 & 3 \end{array}}{\begin{array}{cc} 7 & 7 \end{array}} + \text{(C)}$$

$$\frac{\begin{array}{cc} O & E \\ E & E \end{array}}{\begin{array}{cc} O & O \end{array}} + \text{(B)}$$

$$\frac{\begin{array}{cc} 3 & 4 \\ 3 & 5 \end{array}}{\begin{array}{cc} 6 & 9 \end{array}} + \text{(D)}$$

SOLUTIONS TO PRACTICE THREE

Exercise 1

Dina walked 6 steps up from the bottom of a staircase while Lila walked 5 steps down from the top. They met on the same step. How many steps does the staircase have?

Solution 1

The staircase has 11 steps.

Exercise 2

When Arbax, the Dalmatian, celebrates his 7th birthday, Lila will be 16. Today, it is Arbax's birthday and Lila is 10. How old is Arbax?

Solution 2

Arbax is one year old.

Exercise 3

Dina's mother is cooking stuffed mushrooms. For each mushroom, she chops one tomato and two scallions. She fills a tray with 6 stuffed mushrooms. How many vegetables has she used in total?

Solution 3

She has used 6 tomatoes and 12 scallions. The total number of vegetables she used is:

$$6 + 6 + 12 = 24$$

Exercise 4

There are four chairs at the breakfast counter in Lila's kitchen. At breakfast, Dina, Lila, and their parents sit down for their meal. How many legs can one count?

Solution 4

Four chairs have 16 legs. Four people have 8 legs. There are 24 legs in total.

Exercise 5

On a good spring day, one can see deer and hares from Dina's porch. Today, Dina saw as many hare's ears as deer's tails. If Dina saw 7 hares, how many deer did she see?

Solution 5

If Dina saw 7 hares, then she saw 14 hare's ears. Therefore, there were 14 deer's tails, which means Dina saw 14 deer.

Exercise 6

Lila gave Amira 12 magic wands. Amira gave Dina 8 magnets. Dina gave Lila 14 crystals. Which girl experienced the largest difference in her total number of toys? Was it an increase or a decrease?

Solution 6

Lila received 14 toys and gave away 12 toys. Her toys decreased in number by 2. Dina received 8 toys and gave away 14 toys. Her toys decreased in number by 6. Amira received 12 toys and gave away 8 toys. Her toys increased in number by 4. Dina's toys changed the most in number.

Exercise 7

Arbax, the Dalmatian, had stashed away 25 bones. Some cats discovered his cache and managed to run away with 2 bones each as Arbax chased them away. Arbax is a smart dog who can count. He counted the bones and found that there were 17 left. How many cats were there?

Solution 7

Arbax lost $25 - 17 = 8$ bones. If each cat stole 2 bones, there must have been 4 cats.

Exercise 8

Lila has 18 new blue pens. Amira proposes an exchange: she will give Lila one fluorescent pen for every 7 blue pens. Lila wants 4 fluorescent pens. Amira will accept a scented eraser instead of 2 blue pens to make up for the difference. How many erasers does Lila have to give Amira?

Solution 8

Lila can exchange 14 blue pens for 2 fluorescent pens and will have 4 blue pens left over. Lila needs another 14 blue pens to get 2 more fluorescent pens. She has 4 of these blue pens and is missing another 10, but she can pay Amira in scented erasers. Five scented erasers will make up for the 10 missing blue pens.

Exercise 9

Which two consecutive months have 62 days in total?

Solution 9

There are two solutions:

- July and August.
- December and January.

Exercise 10

Five rats, two squirrels, and one mongoose decide to start a small business. In one day, each rat can bring in 10 clients and each squirrel can bring in 6 clients. The mongoose is in charge of the supply chain and has to provide 4 dollars worth of merchandise per client. How many days will it take the mongoose to spend 128 dollars?

Solution 10

Each day, the total number of clients is $10 + 6 = 16$. If the mongoose spends 4 dollars for each client, the total daily cost is:

$$16 + 16 + 16 + 16 = 64 \text{ dollars}$$

This happens to be the half of 128. It will take them only two days to spend 128 dollars!

Exercise 11

Each day, Penelope weaves 6 feet of fabric. Each night, she unravels 5 feet of fabric. How many days does it take her to wave a 12 ft long coverlet?

Solution 11

Within a 24-hour day, Penelope manages to add one foot of length to her coverlet. If this goes on for 6 days, the coverlet is then 6 feet long. On the seventh day, Penelope weaves 6 more feet to the length and... the coverlet is finished! She needs 7 days.

Exercise 12

An unusual weather pattern started on a Monday. Each day it rained for three hours as follows: the first day it rained from midnight to 3 am, the next day it rained from 3 am to 6 am, and so on. What day of the week was it when the rain started at midnight for the second time?

Solution 12

Draw a circle to represent a clock. Put a tick mark on it every three hours: at 12, 3, 6, and 9. On the outside, write down which day of the week it rained in that interval. After you get to noon, continue during the day until midnight - another 12 hours. This will take you to the next Monday. The rain will start again at midnight on a Tuesday.

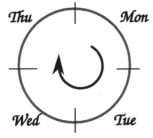 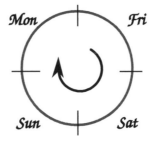 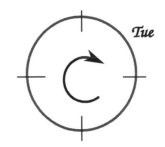

Exercise 13

Five bears foraged together in a forest. They decided to take turns to rest as follows: one of them would rest while the other four foraged. If each bear rested for 6 hours, how many hours did each bear spend foraging?

Solution 13

For each hour of rest, a bear foraged for 4 hours. For 6 hours of rest, the bear foraged for 24 hours.

Exercise 14

It's dance recital day! Lila had her performance at 3 pm and arrived 2 hours ahead of time for rehearsal. Dina had her performance at 2 pm and arrived 3 hours ahead of time, when the hall opened, for rehearsal. Each performance was one hour long and the hall closed after Lila's performance. For how many hours was the hall open?

Solution 14

The hall opened at 11 am, when Dina arrived. Lila performed from 3 pm to 4 pm, after which the hall closed. The hall was open for a total of 5 hours.

Exercise 15

Lila's train, which was supposed to arrive at 3:45 pm, arrived 20 minutes late. As a result, Lila missed the 3:50 bus and had to wait 25 minutes for the next bus. If the bus ride home takes 15 minutes, how late was Lila compared to her usual arrival time?

Solution 15

If the train had been on time, Lila would have taken the 3:50 bus.

However, the train arrived at 4:05 and Lila had to wait 25 minutes for the bus. Lila got on the bus home at 4:30 pm.

Since the bus ride home takes the same amount of time in both cases, it is not necessary to know how long it takes.

The difference in time between 3:50 pm and 4:30 pm is 40 minutes. Therefore, Lila arrived home 40 minutes later than usual.

Exercise 16

Dina has a Book of Magic that is numbered backwards. If she is now on page 201 and the book has 453 pages, how many pages has she actually read?

Solution 16

She has read $453 - 201 = 252$ pages.

Exercise 17

Cornelia, the shepherd, has 70 animals: sheep, hens, and cows. There are as many sheep's legs as hen's legs. There are as many cow's tails as hen's legs. How many cows are there?

Solution 17

Solve this problem by making a diagram. Since a sheep has four legs and a hen has only two legs, there must be more hens than sheep. Make a rectangle to represent the number of sheep:

The number of hens is going to be twice as large:

Hens

Since there are two cows for each pair of hen's legs, there are twice as many cows as hens.

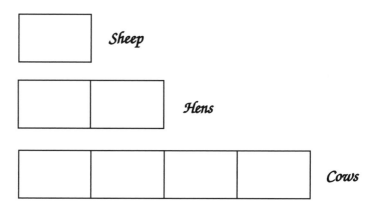

Count the small rectangles. There are 7 of them. If Cornelia has 70 animals, then each rectangle represents 10 animals. There must be 10 sheep, 20 hens, and 40 cows.

Exercise 18

How many pounds of cabbage at 2 dollars a pound weigh the same as 5 pounds of onions at 1 dollar a pound?

Solution 18

Since pounds are used to measures mass (weight), 5 pounds of cabbage will weigh the same as 5 pounds of onions. The other data does not play a role.

Exercise 19

Lila has to help her mother mail some reports. She uses a staple for every 3 sheets of paper. However, out of 100 staples, 20 will break and will need replacement. How many staples does she need if she has to staple 480 sheets in total?

Solution 19

If Lila stapled 3 sheets together in a bundle, then she made 160 bundles out of the 480 sheets. Since 20 staples broke and needed replacement, the total number of staples she used was $160 + 20 = 180$.

Exercise 20

Ali and Baba are exploring the cave of the forty thieves. The entrance is locked, but Baba has a magic key that unlocks any lock. From the entrance, 5 corridors branch out, each of them locked. Each corridor has 3 rooms on one side and 3 rooms on the other side, all locked. How many of the rooms will Ali and Baba be able to explore if the magic key loses its power after 20 uses?

Solution 20

In the diagram, the circular shapes represent doors. Doors opened by using the magic key have been shaded:

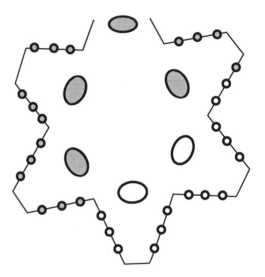

To explore all the rooms in a corridor, the key must be used 7 times. To open the main entrance and explore 2 corridors, Baba will use the key 15 times. Baba will be able to open another corridor and visit 4 of the 6 rooms in there. In total, they will be able to explore $6+6+4 = 16$ rooms.

SOLUTIONS TO MISCELLANEOUS PRACTICE

Exercise 1

Jerome had twelve properly outfitted boats ready to be rented out at his beach sports shop. Each boat is required to have two paddles and four lifesavers. A storm swept away five paddles and eight lifesavers from the shop. How many operational boats did he have remaining?

Solution 1

The loss of five paddles makes three boats unfit for use. The loss of eight lifesavers makes two boats unfit for use. Jerome will assign all the losses to as few boats as possible, leaving only three boats stranded on shore. Jerome will still be able to rent out 9 boats.

Exercise 2

Lila is learning how to cook. She has a recipe for cookies that calls for 2 cups of flour, 3 sticks of butter, one cup of chocolate chips, and three eggs. She uses this recipe and makes 26 cookies. How many cups of flour should she use next time if the wants to bake 39 cookies?

Solution 2

Each cup of flour makes 13 cookies. Three cups of flour will make 39 cookies.

Exercise 3

Dina, Lila, and Amira have 5 plush toys. Can you help them share the toys so that each of them has a different number of toys? (Each girl must have at least one toy.)

Solution 3

No, you cannot help them share the toys. If they each have a different number of toys, the smallest numbers they could have are 1, 2, and 3, for a total of 6 toys. Since they only have 5 toys, two of them will have the same number of toys, regardless of how they share them.

Exercise 4

Ali discovered a box filled with gold coins in the thieves' lair. Ali managed to stuff half of the coins in his pockets and Baba managed to stuff half of the remaining coins in his pockets, before they heard noises from outside and slipped out of the cave. The returning thieves found 30 coins in the box. How many coins did Ali and Baba steal?

Solution 4

Solve this problem by reasoning backwards from the end to the beginning of the story. If there were 30 coins left in the box and Baba had just taken half of the coins, then there were 60 coins in the box before Baba took his share. These 60 coins represented the half left by Ali, who must have taken the other 60 coins. Therefore, there were 120 coins in the box.

Exercise 5

If the letter **O** represents an odd digit, and the letter **E** represents an even digit, which equalities are impossible?

Solution 5

Start your solution to this problem by finding examples that work:

(A) $O + O = E$ $\qquad$ $3 + 5 = 8$

(B) $OO - OO = EE$ $\qquad$ $77 - 55 = 22$

(C) $OO - O = EE$ $\qquad$ $31 - 5 = 26$

(D) $EE + O = EO$ $\qquad$ $82 + 5 = 87$

(E) $E + O = EO$

The last equality is not possible, because the largest carryover from adding two digits is 1. Therefore, if there is a tens digit, it has to be equal to 1, which is not even.

Exercise 6

Dina's grandmother has 5 blue cups, 2 red cups, 3 red saucers, 2 blue saucers, and 2 green saucers. How many cup and saucer pairs can she make for which the color of the saucer does not match the color of the cup?

Solution 6

Represent the saucers by a string of letters:

RRRBBGG

Represent the cups by a string of letters:

BBBBBRR

Since the two green saucers do not match any of the cups, we can use the green saucers in any of the pairs. Assign the 2 blue saucers to the 2 red cups and the 3 red saucers to 3 of the blue cups. Finally, assign the green saucers to the remaining cups. You have obtained seven cup-saucer pairs:

BR, BR, BR, RB, RB, BG, BG

Exercise 7

Tony, the car mechanic, wants to have equal amounts of coolant fluid in two containers. He has 4 liters of coolant in one container and 6 liters in the other. How many liters of coolant should he transfer from the first container into the second one?

Solution 7

Tony has to pour 1 liter of fluid from the container with 6 liters into the other container. Now, there are 5 liters of coolant in each container.

Exercise 8

Dina and Amira are standing in line at the baker's shop. There are four people in front of Dina and four people between Dina and Amira. By the time Dina places her order, Amira will be:

(A) third in line

(B) fourth in line

(C) fifth in line

(D) sixth in line

(E) eighth in line

Solution 8

Dina is first in line when she is ordering. Count 4 more places down the line to reach Amira's position. Amira is 5th in line.

Exercise 9

Lila and Dina are playing "Little Romans." They have to dress in togas, speak Latin, and solve the following:

$$MCM - MC =$$

Solution 9

$$MCM - MC = DCCC$$

Solution 10

Provide results expressed as Roman numerals:

(a) $CM + MC = MM$

(b) $MCCC + DCCC = MMC$

(c) $XI + IX = XX$

(d) $MC - CM = CC$

(e) $XI - IX = II$

(f) $XXIV + XXVI = L$

(g) $XXVI - XXIV = II$

(h) $LX - XL = XX$

(i) $MMM - CCC = MMDCC$

(j) $MCL - CML = CC$

(k) $CXX - LXX = L$

(l) $CCC - XXX = CCLXX$

Exercise 11

1. How many two digit numbers have identical digits?
2. How many three digit numbers have identical digits?
3. How many four digit numbers have identical digits?
4. How many one hundred digit numbers have identical digits?

Solution 11

For each of the questions, there are 9 numbers with identical digits. Such a number is called a *repdigit*.

Exercise 12

Lila has some cards with the digit 6 on them. By turning a card upside down, she can get a card with the digit 9 on it. Lila forms two digit numbers with these cards. What is the sum of all the different numbers she can make?

Solution 12

$$66 + 69 + 96 + 99 = 330$$

Exercise 13

In the figure, the square and the circle represent different digits. What is the difference between the largest and the smallest digit values the triangle can represent?

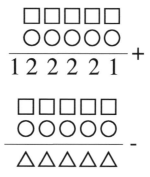

Solution 13

The sum of 122221 can be obtained only if there is a carryover from adding the square and the circle. Because the last digit of the number 12221 is 1, the square and the circle must add up to 11. Since the difference is positive, the square must represent a larger value than the circle. Possible pairs of digits that satisfy are: $6 + 5$, $7 + 4$, $8 + 3$, and $9 + 2$. The possible values of the digit represented by the triangle are: 1, 3, 5, and 7. The difference between the largest and smallest values is $7 - 1 = 6$.

Exercise 14

The figure represents a correct addition:

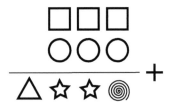

In the figure, different shapes represent different digits and the same shape always represents the same digit. Which of the following values can the spiral not have? Check all that apply.

Solution 14

The presence of the triangle means that the circle and the square must represent digits that add up to 10 or more, because there is a carryover. Therefore, we cannot achieve 9 in the place of the spiral by using additions like: $1+8$, $2+7$, $3+6$, or $4+5$. Since two digits that produce a carryover add up to at most 18, the spiral cannot be 9. Also, because different symbols must represent different digits, the spiral cannot be 0, 7, or 8.

Exercise 15

In the following list, how many numbers are odd?

$$\{4, 10, 151, 200, 0, 45, 736\}$$

Solution 15

4, 10, 200, 0, and 736 are even. 151 and 45 are odd. Therefore, two of the numbers are odd. **Attention!** 0 is even.

Solution 16

1. The largest three digit number with all different odd digits is 975.
2. The smallest three digit number with all different odd digits 135.
3. The largest three digit number with all different even digits 864.
4. The smallest three digit number with all different even digits 204.

Exercise 17

The following sequence is built according to a pattern. Help Dina add three more terms to it!

$$79, \ 69, \ 60, \ 52, \cdots$$

Solution 17

Notice that the difference between neighboring (*adjacent*) terms is a sequence of consecutive numbers:

$$
\begin{aligned}
79 - \mathbf{10} &= 69 \\
69 - \mathbf{9} &= 60 \\
60 - \mathbf{8} &= 52 \\
52 - \mathbf{7} &= 45 \\
44 - \mathbf{6} &= 39 \\
38 - \mathbf{5} &= 34
\end{aligned}
$$

Exercise 18

The sum $\Diamond + \Diamond$, where $\Diamond$ represents a digit, is:

(**A**) odd

(**B**) even

(**C**) impossible to determine

(**D**) neither even nor odd

Solution 18

The result of the operation is twice the $\diamond$. It is, therefore, even. More-over, for answer (D), note that no integer is neither even nor odd. All integers have well defined parity.

Exercise 19

Amira has to make 3 different prizes for the swim meet. She has 5 toys to use as prizes, but each prize has to consist of a different number of toys. Can you explain why she is a bit puzzled?

Solution 19

Even if Amira starts by giving only one toy to one of the prizewinners, she has to give a different number of toys to the next prizewinner. The smallest number of toys she can give the next winner is 2. Similarly, the smallest number of toys she can give the next winner is 3. Amira realizes that, to make three prizes with different numbers of toys, she must have at least 6 toys to distribute.

Exercise 20

The largest of five consecutive odd numbers is 19. How many of these numbers are greater than 14?

Solution 20

The five numbers are 11, 13, 15, 17, and 19. Three of these numbers are greater than 14.

Exercise 21

Which number does the star represent?

$$\bigcirc + \bigcirc + \bigcirc = 60$$

$$\text{(spiral)} + \text{(spiral)} = \bigcirc$$

$$\bigcirc + \text{(spiral)} = \star$$

Solution 21

Since the three circles have a sum of 60, each circle must represent the number 20. Therefore, each spiral must represent 10 and the star must represent $20 + 10 = 30$.

Exercise 22

Dina wrote four consecutive odd numbers on a piece of paper. Arbax played with it and pierced it with his fangs. Now some of the digits are no longer legible. Can you figure out the missing digits?

$$\_\,3\,1\,\_$$

$$\_\,3\,\_\,9$$

$$5\,\_\,\_\,\_$$

$$\_\,\_\,\_\,\_$$

Solution 22

Determine the last digit of the first number. It must be smaller than 9 by 2 - therefore, it is 7. Since the tens digit for the first number is 1, the hundreds and thousands digits will not change within this sequence.

The numbers are: 5317, 5319, 5321, and 5323.

Exercise 23

Lila had 2 blue cubes worth 5 points each, 2 green cubes worth 3 points each, and 2 yellow cubes worth 6 points each. She placed them in a bag and asked Dina to pick 3 cubes without looking.

1. What is the largest odd total score Dina can obtain?
2. What is the lowest even total score Dina can obtain?

Solution 23

To obtain an odd score from 3 cubes, there must be 2 even scores and one odd score, or 3 odd scores. The largest odd score is 17:

$$\text{Odd + Even + Even = Odd}$$
$$5 + 6 + 6 = 17$$
$$\text{Odd + Odd + Odd = Odd}$$
$$5 + 5 + 3 = 13$$

To obtain an even score from 3 cubes, there must be 2 odd scores and one even score. It is not possible to obtain three even scores since there are only two cubes which are worth an even number of points.

$$\text{Even + Odd + Odd = Even}$$
$$6 + 3 + 3 = 12$$

The lowest even score is 12.

Exercise 24

Dina's mother organized a party for the children of her coworkers. She placed some boxes of candy on a table. Each box contained 20 candies. At the end of the party, Lila noticed a funny thing had happened: one box was left untouched, from another only one candy had been removed, from another two candies were missing, from another three candies, and so on. The final box was completely empty. How many boxes were there in total?

Solution 24

There were 21 boxes. 1, 2, 3, $\cdots$, 20 candies had been removed from different boxes. The last box, which was left empty, was the 20th box. Add the unopened box to these 20 boxes to find the total of 21.

Exercise 25

Lila and Amira are selling 40 of their old toys at a yard sale. Instead of selling each toy for 90 cents, they decided the sale would go quicker if they offered a "buy 3 toys and get one toy free" deal. To make the same amount of money in total, how much do they have to sell each toy for? (Assume they sell all the toys in either case.)

Solution 25

If they sell 4 toys for 90 cents each, they make 3 dollars and 60 cents. If they sell 3 toys for the same amount of money, they have to charge 1 dollar and 20 cents for each toy.

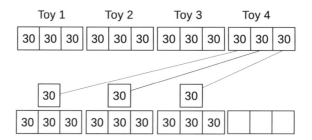

In the figure, we have represented the initial cost of each toy as a sum of the form $30 + 30 + 30 = 90$ cents. In order to be able to give one toy "for free," Lila and Amira charge an extra 30 cents for each of the three toys sold at regular price. The new price is $30 + 30 + 30 + 30 = 120$ cents (1 dollar 20 cents).

Exercise 26

Lila and Dina are trading toys. They agree that one doll should be worth 3 magnets and that one magnet should be worth two UV-beads. Lila has lots of UV-beads. If Lila wants to trade UV-beads for a doll and a magnet from Dina, how many UV-beads should she give Dina?

Solution 26

Use the diagram to find that one doll is worth 3 beads. Therefore, one magnet and one doll are worth $2 + 3 = 5$ beads.

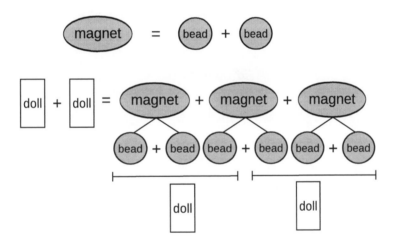

Exercise 27

Amira has stored 759 songs in her collection of favorites. How many more songs does she have to store for the number of songs to be the next number with three different odd digits?

Solution 27

The next number with different odd digits is 791. Amira must store another $791 - 759 = 32$ songs.

Exercise 28

Dina and Lila are allowed to remove 3 digits from the number 13904521 but are not allowed to change the places of the remaining digits. In this way, Dina wants to obtain the largest possible number formed with the remaining digits, while Lila wants to obtain the smallest. What is the difference between Dina's number and Lila's number?

Solution 28

Lila's work:	*Dina's work:*
1 3 9 0 4 5 2 1	1 3 9 0 4 5 2 1
1̸ 3̸ 9̸ 0 4 5 2 1	1̸ 3̸ 9 0̸ 4 5 2 1
4 5 2 1	9 4 5 2 1

The difference is: $94,521 - 4,521 = 90,000$.

Exercise 29

Max, the baker, sells baguettes for 1.20 dollars each and focaccias for 2.30 dollars each. What is the smallest number of baguettes and foccacias he must sell in order to get an amount of money that can be written as whole dollars, no cents.

Solution 29

Two foccacias and one baguette will sell for $3.40 + 3.40 + 1.20 = 8$ dollars. Max has to sell three breads to obtain a whole number of dollars.

Exercise 30

For each puffin that flies off from a cliff, three birds land on the cliff. There were 1000 puffins on the cliff and 20 of them flew off. How many puffins are there now on the cliff?

Solution 30

Each time a bird flies off, the total number of birds on the cliff increases by 2. Since 20 birds flew off, the total number of birds on the cliff increased by 40. There are now 1040 puffins on the cliff.

Competitive Mathematics Series for Gifted Students

Practice Counting (ages 7 to 9)
Practice Logic and Observation (ages 7 to 9)
Practice Arithmetic (ages 7 to 9)
Practice Operations (ages 7 to 9)

Practice Word Problems (ages 9 to 11)
Practice Combinatorics (ages 9 to 11)
Practice Arithmetic(ages 9 to 11)
Practice Operations (ages 9 to 11)

Practice Word Problems (ages 11 to 13)
Practice Combinatorics (ages 11 to 13)
Practice Arithmetic and Number Theory (ages 11 to 13)
Practice Algebra and Operations (ages 11 to 13)
Practice Geometry (ages 11 to 13)

Practice Word Problems (ages 12 to 15)
Practice Algebra and Operations (ages 12 to 15)
Practice Geometry (ages 12 to 15)
Practice Number Theory (ages 12 to 15)
Practice Combinatorics and Probability (ages 12 to 15)

This is a series of practice books. With the exception of a few reminders, there are no theoretical explanations. For lessons, please see the resources indicated below:

Find a set of free lessons in competitive mathematics at www.mathinee.com. Addressing grades 5 through 11, the *Math Essentials* on www.mathinee.com present important concepts in a clear and concise manner and provide tips on their application. The site also hosts over 400 original problems with full solutions for various levels. Selectors enable the user to sort essentials and problems by test or contest targeted as well as by topic and by the earliest grade level they can be used for.

Online problem solving seminars are available at www.goodsofthemind.com. If you found this booklet useful, you will love the live problem solving seminars.

Made in the USA
San Bernardino, CA
19 April 2014